DUTCH AND BELGIAN SHORT SEA SHIPPING 1996-97

by

Roy Fenton

and

Bernard McCall

INTRODUCTION

This book is a natural follow-up to *Short Sea Shipping 1995,* and features one of Europe's largest fleets of short sea ships - that of the Netherlands - with Belgian ships added for good measure. Since the rise of the motor coaster after the first world war, the Dutch have been major players in short sea shipping, and the number and variety of craft listed in this book testifies to their continued success in this highly competitive business.

The Dutch were early advocates of flagging out their coasters, and many of the ships listed here run under flags other than that of the Netherlands. Indeed, this book reflects the increasingly international nature of shipping. For instance, several coasters are included which fly the German flag but are managed from the Netherlands, whilst other ships are operated from Germany but fly the Dutch flag. We have been liberal in our inclusion policy, and have attempted to list all short sea ships which we believe to be owned or managed in the Netherlands and Belgium, although we have omitted ships owned in Germany and elsewhere which use the Netherlands Antilles as a flag of convenience. However, in today's shipping industry, it is quite possible for a ship whose ultimate owner is a citizen of one of these countries to have no apparent links with that country. It should also be noted that published sources frequently do not agree on who is a ship's owner and manager, nor on their location. In these cases, we have generally followed the information given in the publications of Lloyds Register of Shipping. Vessels listed were reported in March 1996 as being in service.

As with all books on short sea or coastal shipping, a somewhat arbitrary upper size limit has had to be applied, and we have decided on 4,000 tons gross. This decision has not been helped by the process of remeasuring tonnage, which has excluded several ships that were recently candidates for this book although for completeness a few just over the upper limit have been admitted. The tonnages quoted are, along with the other details, those reported in March 1996 but may well change further.

The types of ship included are broadly those in *Short Sea Shipping*: dry cargo vessels, tankers and freight ro-ros but not passenger vessels. Within the 4,000 ton limit we have included some of the heavy lift ships and reefers which are Dutch specialities, even though these are not confined to short sea trades.

A few small changes in format have been made for this book. The flag of a vessel is indicated by an abbreviation immediately after its name. These abbreviations, along with those indicating the type of vessel, are listed on page 45. We have revived the custom of giving details of funnel and hull colours where known, as these are often the best way of recognising a company's ships. However, when using these details it should be remembered that ships may carry charterer's colour schemes, and that the managed ships which make up the bulk of some fleets usually carry their individual owner's colours. Hull colours, too, can vary at the owner's or charterer's whim; whilst the colours of modern anti-fouling paints are both difficult to describe and tend to change over time as layers peel off.

For each ship, the following information is given:

SHIP NAME (flag if not Dutch or Belgian) (previous names and year of change)	year built	gross tonnage	deadwt. tonnage	length overall	extreme breadth	loaded draught	service speed	ship type

Over twenty years ago one of the authors was involved in compiling a book which listed Dutch coasters, and it is interesting to look back and find that several ships in that book, *Dutch and German Coaster Fleets,* reappear in the present work. Over the two decades, the change in the individual fleets listed has been considerable, and it is notable how often former coaster owners have moved up to larger ships. There are a few major companies which have disappeared, notably the late-lamented Carebeka, but it is good to see others continuing to prosper, including Wagenborgs who, in their centenary year, have the largest fleet in this book, and Becks who have clung to relatively conventional coasters and have not followed the fashion for low-air-draft vessels.

The authors gratefully acknowledge the help they have received, particularly from Louis Loughran and his international network of observers of funnel colours, from Barbara Jones and Anne Cowne at Lloyds Register of Shipping, from Gil Mayes who undertook a thorough check of our original drafts, from those who supplied photographs, and from Heather Fenton for word and number processing.

We have tried our best to achieve completeness and accuracy, but would be pleased to hear from anyone who has further information or corrections. The authors cannot accept responsibility for any errors or omissions or their consequences.

It is hoped to produce an updated edition of *Dutch and Belgian Short Sea Shipping* every two years. Meanwhile, the authors are turning their eyes towards another large European short sea fleet, that of Germany.

Roy Fenton
(Wimbledon)

Bernard McCall
(Portishead)

DUTCH SHORT SEA SHIPS

AMASUS SHIPPING B.V., Farmsun

Funnel: *Vessels carry owner's funnel markings, but often fly the Amasus houseflag which is red above blue with a large white diamond with the letters S in red and A in blue.*
Hull: *Various, but often blue.*
Agents only for:

V.o.f. J. Bakker & Mevr. Bakker-van der Woude								
ALDEBARAN	84	997	1506	78.9	10.1	3.1	9	gen
(ex Bornrif-94)								
Maatschappij Messchendorp c.s.								
ANTARES	84	1172	1576	78.9	10.1	3.4	9	gen
(ex Oldehove-94)								
V.o.f. Aquatique								
AQUATIQUE	62	393	580	55.0	7.2	2.3	9	gen
(ex Exodus-94)								
Marsani Shipping Co. N.V.								
AURIGA (NA)	78	1589	1670	84.2	10.8	3.7	10	gen (63c)
(ex Algerak-90, Germann-86)								

AURIGA in the Mersey *(Ambuscade Marine Phototgraphy)*

Rederij Chr. Kornet & Zonen B.V.								
CHRISTIAAN	84	1399	2280	79.8	11.0	4.1	9	gen (86c)
(ex Mouna-93)								
EBEN HAEZER	86	1272	1500	81.2	10.4	3.5	10	gen (81c)
V.o.f. Compaen Shipping								
COMPAEN	75	935	1476	80.1	9.0	3.2	10	gen (38c)
(ex Argo-93, Inez V-83, Cargo-Liner V-81)								
C. Veninga & Zn. C.V.								
EEMSHORN	95	2735	4250	89.6	13.8	5.7	12	gen (245c)
H. Ten Napel								
ELISABETH G	78	723	1163	62.5	9.9	3.4	9	gen (33c)
(ex-Vera-95, Frisian-90, Frisiana-89, Thalassa-79)								
Gitana C.V.								
GITANA	70	605	830	67.0	8.1	2.6	11	gen
(ex Aldebaran-93, Gesina-90, Banjaard)								
Maatschappij Ideaal								
IDEAAL	79	726	1163	62.5	9.4	3.2	10	gen
(ex Meran-90, Sambre-89, Carpe Diem II-81)								

Gersom B.V.								
LAURINA NEELTJE	95	1546	2106	84.8	10.8	5.4	12	gen
V.o.f. Miska								
MISKA	63	429	517	55.6	7.3	2.5	10	gen
(ex Rola-90, Corrie II-86)								
J. Bakker								
MORGENSTOND	67	439	530	62.1	7.2	2.2	9	gen
(ex Texel-81)								
V.o.f. Njord								
NJORD	85	998	1490	78.8	10.6	3.1	10	gen
(ex Deo-Juvante-95, Huibertje Jacoba-87)								
Wolderwijd Shipping								
QUO VADIS (NA)	83	1130	1564	79.0	10.0	3.5	10	gen
Scheepvaartonderneming Ulmo								
ULMO	61	279	605	57.3	7.2	2.0	10	gen
(ex Papillon-95, Drakkar-89, Frisiana-79, Alja-M-73)								

A coaster, but looking like an inland craft, ULMO is seen outward bound on the River Elbe *(Bernard McCall)*

C. Leyten								
VESTING	92	1578	2166	88.0	11.9	3.6	10	gen
Vios C.V.								
VIOS	84	998	1485	78.0	10.0	3.2	10	gen
A. Visser								
WARBER								
(ex Zeus-89)	65	455	571	60.5	7.5	2.2	10	gen
Rederij m.s. "Zuiderzee" C.V.								
ZUIDERZEE	85	698	990	63.4	10.0	3.1	11	gen

AMONS & CO., Zaandam

Funnel: *Vessels carry owner's funnel markings.*
Hull: *Various.*
Managers for:

Rederij H & F Bruins C.V.								
HENDRIK-B	82	3210	3760	82.4	15.8	6.0	11	gen (146c)
(ex Polarborg-95)								
Rederij Marathon								
MARATHON (NA)	76	1655	2575	78.7	12.4	5.0	11	gen
(ex Vrouwe Alida-90)								

MARATHON passes Beachley on her way to Sharpness *(Richard Potter)*

Southern Marine Services N.V.

MARINER (NA)	66	1148	1821	73.0	12.1	5.0	11	gen

(ex Armada Mariner-83, Dicky-79, Kraftca-75, Jan F. Spliethoff-66)

Nimbus Marine Ltd.

REIN-B (SV)	58	492	710	48.8	9.0	3.5	9	gen

(ex Hendrik B-93, Karina-90, Karin-89)

ARPA SHIPPING B.V., Roosendaal

Funnel: *Plain blue.*
Hull: *Blue with ARPA in white and red boot topping.*
Managers for:

E.T. Shipping Ltd.

ELIANE TRADER (Ma)	78	1384	1623	81.6	10.0	3.4	10	gen

(ex Konigssee-95)

J.T. Shipping Ltd.

JOHANNA TRADER (At)	83	1980	3009	81.7	14.1	5.4	11	gen (95c)

(ex Vrouwe Johanna-95)

L.T. Shipping Ltd.

LOUISE TRADER (Ma)	81	1716	2319	80.3	11.4	4.3	11	gen

(ex Katharina-94, Selena-89)

M.T. Shipping Ltd.

MICHELLE TRADER (Ma)	83	994	1591	63.7	11.8	3.9	10	gen

(ex Paola-95, Paul Brinkman-88)

S.T. Shipping Ltd.

SAMIN TRADER (Ma)	79	1456	1703	81.9	10.0	3.6	10	gen

(ex Sea Mosel-94, Huberna-89)

ASPEN SHIPPING & TRADING B.V., Amsterdam

Funnel: *Usually black with letter M in white or variants.*
Hull: *Usually black with MUNDIAL LINE in white.*
Managers for:

Joseph Saade & Co.

AWAD I (Le)	66	298	615	47.9	9.2	3.4	12	gen

(ex Juliana-93, Granitz-82)

COSETTE (Le)	66	298	600	47.8	9.2	3.4	12	gen
(ex Sagard-83)								
Mirapolis Shipping Corporation								
BASSMA (Ho)	65	499	1145	66.2	10.6	4.0	11	gen
(ex Arosandra-91, Hannes Lunstedt)								
Mundial Trust S.A.								
FEEDERMATE (Be)	73	3495	1194	92.6	14.8	4.3	14	ro
(ex Tony Car-95, Tony Safi-87, Ramsgate-84)								
Yehla Amine Kabbani								
GENO (Le)	71	299	718	57.7	10.3	3.7	12	gen (25c)
(ex Hagenow-91)								
Esmeralda Maritime Corporation								
KESSMA (Ho)	65	399	1065	61.7	10.0	4.0	11	gen
(ex Ocean Sun-88, Santana-87, Ruthensand-80, Martha Ahrens-75)								
Beliban Car S.a.r.l.								
MUNDIAL CAR (Le)	65	399	2133	81.8	14.0	5.3	13	ro
(ex Passat-82)								
Mundial Line S.a.r.l.								
O'SHEA EXPRESS (Le)	70	398	1054	91.5	16.8	3.9	14	veh
(ex Clearway-78, Speedway-70)								
Also larger vessels.								

EVERT BAYS C.V., Oudeschild

Funnel: *Yellow with black top.*
Hull: *Green with red boot topping.*

WILLY	86	851	1281	64.2	10.5	3.3	10	gen
WILLY II	91	1189	1891	64.7	11.5	4.4	10	gen

Leaving Liverpool is the WILLY II *(Ambuscade Marine Photography)*

BECK SCHEEPVAARTKANTOOR B.V., Groningen

Funnel: *Blue with italic letter B in white inside white ring.*
Hull: *Grey with black boot topping.*
Managers for:

Rederij m.s. "Alert"								
ALERT	84	2970	4830	87.0	15.4	6.7	12	gen
Kustvaartrederij "Arrow"								
ARROW	88	2986	5150	92.1	15.2	6.7	11	gen
B.V. Globecka, B.V. Nobecka & B.V. Tribecka								
COMTESSE	94	3968	6630	100.7	16.0	7.0	12	gen
TRITON	86	997	1544	63.8	11.9	3.9	10	gen
TRIUMPH	86	997	1544	63.8	11.9	3.9	10	gen
VEDETTE	90	2033	3502	86.0	14.2	5.3	12	gen (152c)
Rederij m.s. "Electron"								
ELECTRON	83	1923	3145	80.6	13.6	5.7	12	gen

The sun sets as ELECTRON sails from Rotterdam *(Roy Fenton)*

Rederij m.s. "Noblesse"								
NOBLESSE	80	1095	1637	65.1	11.4	4.4	10	gen
Rederij "Proton"								
PROTON								
(ex Swallow-81)	74	1828	3192	80.4	13.6	5.5	12	gen
Rederij "Valiant"								
VALIANT	77	1893	3098	86.0	14.2	5.3	12	gen
Rederij "Velox"								
VELOX	92	2033	3502	86.0	14.2	5.3	12	gen (152c)
Rederij "Verona"								
VERONA	82	1923	3095	80.0	13.4	5.6	12	gen
Rederij m.s. "Victress"								
VICTRESS	82	1095	1622	66.2	11.6	4.5	11	gen

BRINKMAN MARINE SERVICES B.V., Groningen

Funnel: *Black with red top and intertwined letters U and B in yellow.*
Hull: *Dark blue with red boot topping.*
Managers for:

Galliard Shipping Ltd.

ALLIANCE (Cy)	79	1094	1551	65.8	10.8	4.3	11	gen
(ex Galliard-94, Alliance-87)								

ALLIANCE in the Solent *(Chris Bancroft)*

Hugo Marine Ltd.

HUGO (Cy)	78	1094	1554	65.8	10.8	4.3	11	gen
(ex Hugo Brinkman-87)								

Paulina B Shipping Ltd.

PAULINA B (Cy)	75	1096	1589	65.8	10.8	4.3	10	gen
(ex Paulina Brinkman-87)								

BROERE SHIPPING B.V., Dordrecht

Funnel: *Black with blue band carrying letters G B in white.*
Hull: *Red.*

DUTCH PILOT	84	2137	3160	91.2	13.7	5.1	12	ch tk
(ex Jacqueline Broere-93)								
DUTCH PROGRESS	84	2137	3160	91.1	13.7	5.1	12	ch tk
(ex Neeltje Broere-93)								
ENGELINA BROERE	75	1807	2610	82.4	12.7	5.7	13	ch tk
(ex Chemtrans Arcturus-80)								

Managers for:

Pakhoed Shipping Ltd

BASTIAAN BROERE	88	3693	5098	104.3	17.0	6.2	15	ch tk
DUTCH GLORY	75	1640	2321	80.2	12.1	5.4	13	ch tk

DUTCH MASTER	75	1640	2321	80.2	12.1	5.4	13	ch tk
DUTCH MATE	89	3693	5098	104.3	17.1	6.2	15	ch tk
DUTCH NAVIGATOR	91	3693	5098	104.3	17.1	6.2	15	ch tk
Dutch Engineer B.V.								
DUTCH ENGINEER	86	2183	2570	80.9	14.5	5.1	14	ch tk
Dutch Mariner B.V.								
DUTCH MARINER	86	2183	2570	81.0	14.5	5.2	14	ch tk
Dutch Sailor B.V.								
DUTCH SAILOR	81	3427	4387	91.0	16.0	6.3	14	ch tk
(ex Broere Aquamarine-87)								
Tankvaart Dordrecht B.V.								
JACOBUS BROERE	89	3693	5098	104.3	17.0	6.2	15	ch tk
New buildings:								
DUTCH CHIEF	96	4000	4450					ch tk
DUTCH FAITH	96	4000	4450					ch tk
DUTCH SPIRIT	96	4000	4450					ch tk

The chemical tanker DUTCH MASTER sails from the River Tees *(Richard Potter)*

HENDRIK BUMA, Nijemirdum

Funnel: *Blue with white band carrying red heart device.*

Managers for:

Kosmima Shipping Ltd.								
BRANDARIS (Cy)	72	1132	1109	68.3	10.5	2.9	11	gen
(ex Osenberg-88)								

CAPELLE CHARTERING & TRADING B.V., Capelle a/d Ijssel

Funnel: *Black with white band with blue device consisting of interlocked angular letter C and reversed C.*

Managers for:

Leandros Shipping Ltd.								
ALECTO (Cy)	84	3259	5050	91.9	15.1	6.5	12	gen
(ex Clarissa-94)								
Flevo Shipping Ltd.								
ALICE (Cy)	77	1104	1580	65.7	10.8	4.3	12	gen
(ex Calypso-95)								

Elina Shipping Ltd.

ELINA (Cy)	75	1010	1559	65.8	10.8	4.3	11	gen
(ex Elina B-90, Elisabeth Holwerda-87)								

Navy Sum Shipping Ltd.

IRENE 1(Cy)	78	1010	1548	65.7	11.1	4.3	12	gen
(ex Irene-95, Azolla-93)								

CEBO MARINE B.V., Heemstede

Funnel: *Blue with white device consisting of letters c b and drilling rig.*
Hull: *Grey with red or black boot topping.*

CARAVELLE	72	999	1270	70.6	10.4	3.5	12	cem
(ex Geeststroom-76)								
CARINE	69	999	1280	70.6	10.4	4.0	12	cem
(ex Bart-83, Garza-82, Geestduin-76)								

CARINE is outward bound from Goole having brought a cargo of cement from Aalborg in Denmark *(David H. Smith)*

GASTANKVAARTMAATSCHAPPIJ CHEMGAS B.V., Rotterdam

Funnel: *White; some have blue panel carrying red houseflag which has white diamond bearing black outline hexagon.*
Hull: *Blue red boot topping; some with CHEMGAS HOLLAND in white.*

TROUT	90	1997	1520	105.6	11.9	2.9	12	lpg
ZEPHYR	85	1621	1296	92.6	11.4	3.2	11	lpg

Managers for:

Chemgas Intercoastal C.V.

SALMON	87	1605	1294	92.8	11.4	2.9	10	lpg
STURGEON	88	1605	1294	92.5	11.4	2.9	11	lpg
TWAITE	91	1997	1640	105.6	12.0	3.0	11	lpg

DAMMERS & VAN DER HEIDE'S SHIPPING & TRADING Co. B.V., Groningen

Funnel: *White with blue band or plain blue, both with orange letter D.*
Hull: *White with red boot topping.*
Managers for:

N.V. Shipping Co. Blanca

CASABLANCA (NA)	80	2989	3536	101.9	14.5	5.3	14	ref

Managed by Dammers Shipmanagement B.V., Willemstad:

Arcturia Navigation Co. Ltd.

ARCTIC (Bs)	83	3526	4156	99.0	16.0	6.7	15	ref
(ex Sapporo Maru-89)								

B.B. Nautic Shipping Co. Ltd.

NAUTIC (Bs)	83	3955	5336	108.5	16.3	7.4	16	ref (93c)

B.B. Nordic Shipping Co. Ltd.

NORDIC (Bs)	84	3955	5265	108.8	16.3	7.4	16	ref (93c)

C.V. SCHEEPVAARTONDERNEMING DIAMANT, Delfzijl

Funnel: *White with letters M in red and D in blue.*

DIAMANT	85	998	1497	78.0	10.0	3.2	10	gen
(ex Watum-95)								

REDERIJ P. DOORENWEERD

UNION ROBIN	83	1525	2310	78.9	12.1	5.1	11	gen
(ex Elisabeth-S-95)								

This vessel is managed by Union Transport Group plc, Bromley, UK

GENCHART V.o.f. (General Shipping and Chartering Services), Rotterdam

Funnel: *Yellow with blue letter G. Managed ships: blue with black top and blue letter G on white panel.*
Hull: *Blue with black or red boot topping.*

STEEL SHUTTLE	85	993	1715	64.8	11.1	4.3	10	gen
STEEL SPRINTER	85	993	1715	64.8	11.1	4.3	10	gen

Living up to her name, STEEL SPRINTER hurries down the River Trent *(Richard Potter)*

Managed by the associated H.S.S. Holland Ship Service B.V.:

Omniseas Shipping Co. Ltd.

ALIDON (Cy)	78	3702	6110	83.7	17.1	8.4	11	gen
(ex Alida Smits-92)								
Grade Price Shipping Co.								
ALSYDON (Cy)	79	3702	6110	83.7	17.1	8.4	11	gen
(ex Alsyta Smits-93)								
Jettrade Shipping Co. Ltd.								
ANDREALON (Cy)	78	3702	6110	83.7	17.1	8.4	11	gen
(ex Andrea Smits-88)								
Topaz Seal Shipping Co. Ltd.								
BLUE TOPAZ (Cy)	75	900	1473	65.4	10.8	4.1	11	gen
(ex Frisian-87)								
Carol Mary Shipping Co. Ltd.								
CECILIA I (Cy)	82	3692	6110	84.2	17.1	8.4	11	gen
(ex Cecilia Smits-89)								
Carol Linda Shipping Co. Ltd.								
CHRISTINA I (Cy)	82	3692	6110	84.2	17.1	8.3	11	gen
(ex Christina Smits-88)								
Claudia Navigation Co. Ltd.								
CLAUDIA I (Cy)	81	3692	5900	84.2	17.1	8.4	11	gen
(ex Claudia Smits-88)								

FIELD SHIPPING & CHARTERING B.V., Rotterdam

RULEWAVE WARRIOR (SV)	78	1307	1426	84.9	9.5	3.3	10	gen(69c)
(ex Topaz-95, Aramon-94, Markab-86)								
DARSSER ORT (Bz)	76	574	696	49.7	8.3	3.5	10	gen

Managed by Field Ship Management Ltd, Woking,UK

H. & G. TRANSPORT B.V., Beverwijk

Managers for:

Sunset Shipping Co.

FENLAND (Bs)	79	2004	2710	90.8	12.1	5.2	13	ref
(ex Norbrit Vries-88, Boston Sea Lance-83)								

The hatch covers are hosed down as CASPIC leaves the Nieuwe Waterweg *(Jan van der Klooster)*

HANNO-OAM SHIPPING B.V., Rotterdam

Funnel: *Blue with white letters HOS below white device signifying heaps of bulk cargo.*
Hull: *Black, blue or grey with red boot topping.*
Managers for:

Bengalen Shipping Co. N.V.								
BENGALEN (NA)	78	1600	3265	81.7	14.1	5.5	12	gen
(ex Zomerhof-93, Sylvia Omega-85)								
Shipping Co. Caspic N.V.								
CASPIC (NA)	78	1600	3265	81.7	14.1	5.5	12	gen
(ex Vijverhof-93, Sylvia Gamma-85)								
Rederij Cemile B.V.								
CEMILE	91	2370	4270	88.3	13.2	5.7	12	gen
(ex Sandfirden-91)								
Rederij "Combi Trader" C.V.								
COMBI TRADER	75	1399	2807	71.5	13.0	5.7	12	gen (64c)
(ex Ocean Coast-75)								
Rederij Ikiena B.V.								
IKIENA	93	2735	4266	89.6	13.2	5.7	12	gen (190c)
Rederij Katja B.V.								
KATJA	93	2753	4250	89.6	13.2	5.7	12	gen
Carib Sun N.V.								
MAGDALENA	90	2371	4247	88.3	13.2	5.5	12	gen
(ex Kirsten-90)								
Rederij Maria N.V.								
MARJA	93	2715	4293	89.8	13.2	5.5	11	gen
Rederij Marjolein B.V.								
MARJOLEIN (NA)	94	2715	4293	89.8	13.2	5.6	12	gen
Faerder Transport N.V.								
NORTH SEA (NA)	78	1600	3214	81.7	14.1	5.5	12	gen
(ex Elise-92, Sylvia Beta-85)								
Shipping Co. Scotia N.V.								
SCOTIA (NA)	77	1599	3214	81.7	14.1	5.5	12	gen
(ex Vredehof-93, Katja-91, Carib Bird-88, Reestland-87, Sylvia Alpha-85)								

HOLWERDA SHIPMANAGEMENT B.V., Heerenveen

Funnel: *Blue with white letter H in the shape of a two-way arrow with two wavy lines in blue across the centre.*
Hull: *Blue with red boot topping.*

Scheepvaartmij Christiania B.V.								
BUMI RAYA (Pa)	77	3971	4640	98.9	16.2	6.0	13	gen (130c)
(ex Schoterland-92, Laima-92, Schoterland-91, Frisian Liner-86, Jan Tavenier-81)								
Frisian Shiptrade B.V.								
ELSA	85	3986	6025	106.6	18.1	6.6	14	gen (419c)
(ex Frisian Glory-93, Umag Cayenne-93, Clement-92, Frisian Glory-86, Samsun Glory-85)								
Scheepvaartonderneming Fenja C.V.								
FENJA	85	3727	4100	103.5	16.5	8.0	13	gen (341c)
(ex Meteor-95, Maersk Tinto-91, Uwe Kahrs-90, Maersk Tinto-90, Gracechurch Gem-88, Uwe Kahrs-86)								
Scheepvaartonderneming Freya B.V.								
FREYA H (Pa)	78	1886	3001	81.0	14.2	5.3	13	gen (92c)
(ex Bumi Jaya-96, Freya-92, Mizar-89, Aned-88, Gera Holwerda-87, Samsun Express-84, Sundsviken-80, Gera Holwerda-78)								
Scheepvaartonderneming Haskerland								
HASKERLAND	82	4047	6081	92.5	16.2	8.1	12	gen (249c)
(ex Inara-92, Haskerland-91, Frisian Carrier-86, Samsun Carrier-85)								

Also larger ships.

JACZON REDERIJ & HARINGHANDEL B.V., Scheveningen

Funnel: Blue with white star and black top.
Managers for :

Jaczon "Holland Klipper" B.V.								
HOLLAND KLIPPER	89	3999	5363	107.7	16.2	7.6	19	ref
Jaczon "Royal Klipper" B.V.								
ROYAL KLIPPER	87	3999	5000	107.4	16.2	7.5	18	ref

Also larger ships

JAN WIND SHIPPING, Nansum

Managers for:

Laserzone Shipping Ltd.								
LIDA (Cy)	74	992	1448	62.8	10.8	4.1	10	gen

(ex Vissersbank-90, Spray-86, Arina Holwerda-81)

SCHEEPVAARTBEDRIJF JOOSTEN B.V., Hendrik Ido Ambacht

Managers for :

Clearwater B.V.								
CLEARWATER	95	1200	4000	79.9	10.9	4.4	11	ch tk

JUMBO SHIPPING CO. S.A., Geneva

Funnel: *White with houseflag as band. This is divided horizontally green/white/green, with a red elephant on a white disc in the centre flanked by blue and red stars.*
Hull: *Blue with red boot topping.*
Agents only for:

Fairlift N.V.								
FAIRLANE (NA)	77	3481	4415	98.0	17.0	6.0	13	hl
Jumbo Scheepvaart Maats. (Curacao) N.V.								
MIRABELLA (NA)	77	3482	4415	98.0	17.0	6.0	12	hl
Jumbo Navigation B.V.								
STELLAMARE (NA)	82	1496	2850	88.2	15.6	5.6	12	hl

(ex Valkenswaard-87)
Also larger ships

KEES DOUMA BEDRIJFSADVIESBUREAU, Loppersum

Funnel: *Ships carry owner's or charterer's funnel.*
Hull: *Various.*
Managers for:

Dioli Shipping N.V.								
DIOLI (NA)	83	499	1760	82.5	11.4	3.6	10	gen (48c)
(ex Baltica-94)								
C.V. m.v. Duiveland								
DUIVELAND	83	998	1463	78.9	10.1	3.2	12	gen
V.o.f. Noorderling								
NOORDERLING	73	967	1400	80.0	9.0	3.0	10	gen

(ex Tasman-92, Bielefeld-84, Cargo-Liner I-81)

DUIVELAND at Newport *(Danny Lynch)*

R. H. KIRK SCHEEPVAARTBEDRIJF, Enschede

JOHANNA (Bs)	84	1280	1558	74.9	10.6	3.4	10	gen
(ex Sea Trent-95, Echo Venture-90, Sea Trent-90)								

KNSM-KROONBURGH B.V., Rotterdam

Funnel: *Black with two broadly separated white bands.*
Hull: *Black.*

TARAS	76	2219	2560	81.4	13.5	5.0	13	gen (150c)
(ex Ikaria-86)								

Also larger ships

KUSTVAARTBEDRIJF MOERMAN B.V., Rotterdam

Funnel: *Yellow with green/white/green bands broken by letter M in green on white disc. Managed ships may carry variants. Jannicke companies: pale blue with black top.*
Hull: *Red with black boot topping.*

Managers for:

Jannicke Three Shipping Ltd.								
FERRO (Cy)	91	1986	3504	88.2	14.2	5.0	12	gen (104c)
C.V. Arctic Marine								
FORTE	89	3998	4001	90.8	16.2	6.4	15	ro/gen
LARGO	90	3998	4000	90.8	16.2	6.4	15	ro/gen
Rederij Achthaven B.V.								
JUPITER	85	1839	2167	78.6	12.7	4.3	12	gen (124c)
(ex Dirk-86)								
Amstelstraat Management Co. B.V.								
LIBRA	80	2201	2200	79.8	12.8	4.6	12	gen (145c)
(ex Allgard-89, Melton Challenger-88)								

Seen in the Nieuwe Waterweg, ROELOF carries a variant of the funnel colours of Kustvaartbedrijf Moerman B.V. *(Jan van der Klooster)*

J. van Urk junior C.V.								
ROELOF	83	1599	3152	81.6	14.1	5.4	11	gen
(ex Viking-92)								
Rederij Salvinia								
SALVINIA	86	1986	2798	79.8	12.1	4.7	12	gen (128c)
(ex Alblas-94)								
B.V. Tweehaven								
SWIFT	93	2825	4148	90.5	13.4	5.6	11	gen
Jannicke Two Shipping Ltd.								
TINNO (Cy)	91	1986	3504	88.2	14.2	5.0	12	gen (103c)
Jannicke One Shipping Ltd.								
TORPO (Cy)	90	1986	3504	88.2	14.2	4.9	12	gen

G. MOS, Delfzijl

Funnel: *Blue with white letter M.*
Hull: *Blue with red boot topping.*
Agents only for:

Mariship Co. N.V.								
ZWANET (NA)*	77	2029	3100	82.3	13.9	5.3	12	gen

(ex Arpa Sun-92, Bottenviken-88, Regulus-86, Bottenviken-83, Deltasee-77)
* Under arrest at Vigo, Spain since 20 August 1994.

EWALD MULLER & Co. Gm.b.H., Hamburg

Funnel: *Various.*
Hull: *Various.*
Managers for:

Anga Shipping N.V.								
ANGA (NA)	81	499	1766	82.5	11.3	3.6	10	gen (48c)
(ex Vineta-94)								
Seareward Shipping Ltd.								
GERMA (Cy)	78	906	1485	63.0	10.0	3.8	11	gen
(ex Edgerma-94)								

W.L. Mastenbroek								
MEANDER	73	804	1213	76.4	8.1	3.0	10	gen
(ex Latona-95, Balgzand-84)								
Scheepvaartbedrijf De Haan								
RIO-Y-MAR	73	504	721	60.4	7.9	2.4	9	gen
J. de Jonge								
SCOUT MARIN	83	299	1063	74.6	9.5	2.9	10	gen (66c)
(ex Wilke-93, Sea Dart-89, Wilke-88)								

NOORDWEST SHIPMANAGEMENT B.V., Tynaario

Managers for:

Hallvard Shipping Co N.V.								
WILLEM (Bs)	79	787	1223	64.5	10.1	3.4	10	gen
(ex Lenie-95)								

NORFOLK LINE B.V., Scheveningen

Funnel: *Black with blue band bearing seven-pointed white star.*
Hull: *Light blue with red boot topping.*

MAERSK FLANDERS	78	7199	3523	122.9	21.0	4.8	16	ro (90u)
(ex Duke of Flanders-90, Romira-86, Admiral Atlantic-84)								
New buildings								
MAERSK EXPORTER	96			142.0	23.0	5.4		ro(120u)
MAERSK	96			142.0	23.0	5.4		ro(120u)

Managed by The Maersk Co. Ltd., London.

OOST ATLANTIC LIJN B.V., Rotterdam

Funnel: *Red with white over green bands bearing letters OAL in red.*
Hull: *Grey with red boot topping.*
Managers for:

Atlantic Horizon Shipping Corp.								
ATLANTIC COAST (Bs)	77	1943	3124	81.7	14.2	5.4	13	gen
(ex Fivel-81)								
ATLANTIC MOON (Bs)	76	3339	2999	96.5	16.0	5.7	12	gen (90c)
(ex Polydorus-88, Mercandian Admiral-79)								
Adamant Marine Ltd.								
ATLANTIC COMET (Cy)	72	1465	1971	82.1	12.2	4.5	13	gen
(ex Cornelia Bosma-78)								
ATLANTIC RIVER (Cy)	71	1449	2489	77.2	13.0	4.8	13	gen
(ex Munte-81, Norimo-74)								

Also larger ships.

ATLANTIC COAST in the River Ouse *(Richard Potter)*

B.V. SCHEEPVAARTBEDRIJF POSEIDON, Delfzijl

Funnel: *Vessels carry owner's funnel markings.*
Hull: *Various.*
Managers for:

Madora II B.V.								
ADMIRAL	91	4059	5850	108.8	16.6	6.5	15	gen
(ex Sea Admiral-95, Admiraal-92)								
H.P. Lanser								
AFHANKELIJK	74	597	826	62.1	9.1	2.6	11	gen
(ex Ameland-88)								
Rederij "Almenum"								
ALMENUM	92	1425	1830	74.0	11.6	4.4	10	gen
Rederij Aquarius								
AQUARIUS	96	1996	3250	84.8	12.5	6.2	11	gen

AFHANKELIJK passes through the Kiel Canal bound for Szczecin *(Dominic McCall)*

Rederij Ariel C.V.								
ARIEL (SV)	85	978	1411	70.0	10.4	3.5	10	gen
(ex Vita Nova-92)								
J. & D. Damhof								
BARON	75	1746	2790	78.4	12.4	5.4	10	gen
(ex Barok-95, Audrey Johanna-81)								
H. Lubbinge								
CUM DEO	63	634	780	67.1	8.3	2.2	9	gen
(ex Via Maris-87)								
Rederij Daniel								
DANIEL	96	1990	3400	89.9	11.9	-	11	gen (106c)
Rederij "Empire" C.V.								
EMPIRE	91	1981	3265	90.0	12.0	4.9	12	gen
Rederij Zodiac								
FAST WIL	83	1394	2284	79.8	11.1	4.1	9	gen (86c)
(ex Breehorn-93, Banjaard-92)								
C.V. Scheepvaartbedrijf Scherpenisse								
GERARDA	94	2650	4200	89.6	13.1	5.7	12	gen (96c)
Rederij S.J. Switijnk								
HARNS	94	1909	2467	89.2	11.4	4.1	12	gen (66c)
Hoop Navigation Ltd.								
HOOP (Cy)	78	1691	2555	78.7	12.5	5.0	11	gen
Fairstar Shipping Ltd.								
KIM (Cy)	76	1469	2440	81.2	11.9	4.7	11	gen
(ex Imke-87)								

Arriving from Santander, KIM sails up a wintry River Trent *(G. V. Smith)*

Rederij m.v. "Marianne"								
MARIANNE	86	1318	1613	79.1	11.4	3.3	10	gen
(ex Harns-94)								
E & E Shipping V.o.f.								
MARINIER	86	1630	2525	87.9	11.0	4.1	10	gen (140c)
V.o.f. Nasvier								
MAYA EVITA (SV)	66	462	615	55.3	8.7	2.9	10	gen
(ex Maartje-93, Anna Maria-90, Este-80, Gisela Bartels-75)								
Coastal Sky Shipping Ltd.								
MERAK (Cy)	76	1472	2340	76.4	12.0	4.8	10	gen
(ex Geziena-88, Meran-80, Merak-80)								

Jan de Koning Gans Jr.								
OOSTZEE	78	815	1150	63.0	9.5	3.4	10	gen
(ex Zwartewater-95, Almenum I-92, Almenum-92)								
Rederij Seabreeze								
SEABREEZE (NA)	95	1996	3246	84.8	12.4	4.8	10	gen
Tell Shipping Ltd.								
TELL (Cy)	75	1011	1590	65.8	10.8	4.3	10	gen
(ex Ketelmeer-87)								
Troubador B.V.								
TROUBADOUR	92	1789	2450	90.0	11.4	3.9	10	gen (98c)
F.S. Switijnk								
VERITAS	93	1592	2270	82.0	12.4	3.9	11	gen
A.G. Switijnk								
VLIELAND	94	1937	3150	82.4	12.4	4.8	12	gen
Zevenster B.V.								
WADDENZEE	85	1861	3035	91.0	11.4	4.9	10	gen (67c)
Jan de Koning Gans Jr.								
ZWARTEWATER	85	1132	1528	79.3	10.4	3.2	10	gen
(ex Marina-95, Verita-92)								

POT SCHEEPVAARTBEDRIJF, Delfzijl

Funnel: *Green base separated from black top by white band bearing the letters P in black and G in green.*
Hull: *Grey with black boot topping.*

DOGGERSBANK*	95	2785	4171	90.3	13.6	7.2	14	gen
SKAGENBANK*	91	1999	3015	82.2	12.5	4.9	11	gen (128c)
VARNEBANK*	88	851	1280	64.2	10.5	4.0	8	gen
VIKINGBANK†	78	1596	3040	81.7	14.3	5.4	13	gen
VISSERSBANK*	94	1682	2503	81.7	11.1	4.5	11	gen

* Managed by Wagenborg Shipping B.V.
† Managed by Amasus Shipping B.V.

VIKINGBANK leaves Seaham *(Matthew Hunt)*

ROMI SHIPPING & TRADING, Edam

Funnel: *None.*
Hull: *Black and white.*

ALBATROS	99	125	142	31.4	6.1	1.9	7	aux
(ex De Albatros-89, Albatros-88)								

ROYAL SHIPPING B.V., Groningen

Funnel: *Vessels carry owner's funnel markings.*
Hull: *Various.*
Managers for:

m.s. Elan Scheepvaartbedrijf								
ELAN	82	1054	1468	78.9	10.0	3.2	9	gen
(ex Vlieland-93)								
Rederij Harma								
HARMA	79	999	1450	65.0	10.7	4.0	10	gen
Jenema Shipping C.V.								
OSIRIS	85	1163	1685	73.7	11.7	3.7	10	gen
D.S. Shipping C.V.								
RACHEL	84	1162	1685	73.7	11.7	3.7	10	gen
(ex Sunergon-94)								

HARMA passes beneath the Humber Bridge with a full deck cargo of timber *(Richard Potter)*

SANARA B.V., Rotterdam

Funnel: *White with stylised letter S in red.*
Hull: *Blue with red boot topping.*
Agents only for:

Diamond Shipping N.V.								
DIAMOND (NA)	85	1487	2475	79.5	11.4	4.4	10	gen
Silver Pride Shipping N.V.								
RUBY (NA)	86	1512	2475	79.6	11.3	4.4	9	gen

R. SCHURINK, KRIMPEN a/d IJSSEL

Funnel: *Orange*

Rederij Sirenia								
DOLFIJN	89	1987	3000	81.2	12.4	5.0	11	gen (154c)

SEATRADE GRONINGEN B.V., Groningen

Funnel: *Blue with black top and orange flag bearing stylised letter S in white and G in black.*
Hull: *White with red boot topping, usually with SEATRADE in black.*
Managers for:

Cape Vincente Shipping Co. Ltd.								
ADRIATIC (At)	84	3505	4173	99.0	16.0	6.7	15	ref
(ex White Reefer-90)								
Legion Shipping Co.								
ANTARCTIC (Bs)	82	3520	4192	99.0	16.0	6.7	15	ref
(ex Ena Maru-89)								
Atlantic Sun Shipping N.V.								
ATLANTIC ICE (NA)	78	1198	2448	81.0	13.2	5.1	12	ref
(ex Atlantic-89)								
Botnic Sun Shipping N.V.								
BALTIC ICE (NA)	79	2285	2468	81.7	13.2	5.1	12	ref
(ex Baltic-88)								
Celtic Sun Shipping N.V.								
CELTIC ICE (NA)	79	1198	2468	81.7	13.2	5.1	12	ref
(ex Celtic-89)								
Mathilda Shipping Co. N.V.								
JOINT FROST (NA)	79	2595	2854	83.4	14.5	6.0	14	ref (77c)
N.V. Shipping Co. Magdalena								
MAGDALENA (NA)	79	2989	3529	101.9	14.5	5.3	14	ref (60c)
Maya Shipping N.V.								
MAYA (NA)	78	2989	3800	101.9	14.5	5.3	14	ref (60c)
C.V. Scheepvaartonderneming Neerlandic								
NEERLANDIC (NA)	85	3955	5232	108.8	16.3	7.3	16	ref (93c)
Normandic Shipping Co. N.V.								
NORMANDIC (NA)	83	3960	5220	108.3	16.3	7.4	16	ref (93c)
Nyantic Shipping Co. N.V.								
NYANTIC (NA)	84	3970	5336	108.5	16.3	7.4	16	ref (93c)
Oceanic Sun Shipping N.V.								
OCEANIC ICE (NA)	77	2244	2404	82.7	13.6	5.0	13	ref
(ex Oceanic-88)								
Swalan Shipping Co. N.V.								
SWALAN (NA)	78	2980	3800	101.9	14.5	5.3	14	ref
(ex Laura Christina-90)								

SOETERMEER, FEKKES' CARGADOORSKANTOOR B.V., Rotterdam

Funnel: *Vessels carry owner's funnel markings.*
Hull: *Various.*
Managers for:

Barkmeijer Beheer B.V.								
HESTER	79	639	965	59.5	9.9	3.2	10	gen
(ex Gersom-94, Sprinter-93, Gersom-86)								

Beck's VICTRESS inward bound on the River Trent *(Richard Potter)*

Photographed in the River Medway, SWIFT is managed by Kustvaartbedrijf Moerman B.V. *(Bernard McCall)*

Owned by Oost Atlantic Lijn, the Cypriot-flag ATLANTIC COMET is seen in the Solent *(Brian Ralfs)*

EEMSBORG arrives at Stormont Wharf, Belfast *(Alan Geddes)*

The Wagenborg-managed LIBELLE leaves the Herdman Channel at Belfast *(Alan Geddes)*

The German-flag, Dutch-managed SEA MAGULA passes Reedness *(David H. Smith)*

ISABEL managed by Wijnne & Barends B.V. in the River Ouse *(Richard Potter)*

MAYA EVITA in the Kiel Canal *(Bernard McCall)*

Saligot Shipping Services Co. Ltd.								
SPRINTER (Cy)	94	1983	3204	90.4	12.5	4.6	11	gen (116c)
V.o.f. Rederij Bruins & Co.								
TERTIUS	95	2051	3280	89.9	12.5	4.7	11	gen (116c)

SPLIETHOFF'S BEVRACHTINGSKANTOOR B.V., Amsterdam

Funnel: *Orange with black top and houseflag which is quartered diagonally red, orange, blue and white and carrying letter S in black.*
Hull: *Brown with green boot topping.*
Managers for:

C.V. Scheepv. Onderneming "Bakengracht"								
BAKENGRACHT	81	3433	3489	80.2	16.1	6.0	12	gen (165c)
BATAAFGRACHT	81	3433	3489	80.2	16.1	6.0	12	gen (165c)
C.V. Scheepv. Onderneming "Brouwersgracht"								
BARENTZGRACHT	81	3433	3444	80.2	16.1	6.0	12	gen (165c)
BICKERSGRACHT	81	3433	3488	80.2	16.1	6.0	12	gen (165c)
BLOEMGRACHT	81	3433	3444	80.2	16.1	6.0	12	gen (165c)
BROUWERSGRACHT	80	3433	3445	80.2	16.1	6.0	12	gen (165c)
C.V. Scheepv. Onderneming "Groesbeek"								
BEURSGRACHT	81	3433	3448	80.2	16.1	6.0	12	gen (165c)
C.V. Scheepv. Onderneming "Sambeek"								
BONTEGRACHT	81	3343	3448	80.2	16.1	6.0	12	gen (165c)
C.V. Scheepv. Onderneming "Houtmangracht"								
HEEMSKERKSGRACHT	82	2145	4553	97.2	16.1	5.9	12	gen (320c)

Also larger ships.

REDERIJ H. STEENSTRA, Geenemuiden

Funnel: *White with shield in red with S device in white.*
Hull: *Blue with black boot-topping.*

ANNE.S	86	1139	1601	79.1	10.4	3.3	11	gen (80c)
DOUWE.S	87	1311	1771	79.7	11.2	3.7	10	gen (78c)

(ex Torpe-93)
Managed by Union Transport Group plc, Bromley, U.K.

ANNE.S makes one of her regular calls at Par to load china clay for Antwerp *(Bernard McCall)*

SWINSHIP MANAGEMENT B.V., Ridderkerk

Funnel: *Various, but often plain red.*
Hull: *Various, often grey with black boot topping.*

Managers for:								
Alligator Ltd.								
AVOSET (SV)	76	1086	1553	65.8	10.8	4.3	11	gen
(ex Pavonis-92)								
Bareboat charter:								
Onesimus Dorey (Shipowners) Ltd.								
TEAL 1 (SV)	SV	891	1400	60.9	9.8	4.3	11	gen
(ex Hoxa Sound-94, Murell-88)								
Onesimus Dorey (Shipowners) Ltd/Wardwood Chartering Ltd								
EAST MED (Bs)	74	1135	1643	67.7	11.8	4.2	11	gen
(ex Cormorant I-95, Betty C-93, Kerry M-87, Sentence-85)								
NORTH MED (Bs)	78	1181	1690	68.0	11.6	4.3	11	gen
(ex Sea Eagle-I-96, Eileen C-93, Tuskar Rock-90)								
SOUTH MED (Bs)	79	1181	1690	68.0	11.8	4.3	11	gen
(ex Albatross I-95, Janet C-93, Fastnet Rock-90)								
Managers for:								
Cormorant Shipping & Trading Ltd.								
EGRET (SV)	66	633	738	57.4	9.1	2.9	10	gen
(ex Cormorant-86, Moon Trader-86, A. Held-79)								
Medtrans Ltd.								
GANNET (SV)	62	404	690	49.5	8.8	3.1	9	gen
(ex Pax I-90, Pax-84, Stephenson-78, Harma-77, Oster Till-73)								
Artemesia Ltd.								
MARLIN I (SV)	77	721	1010	60.0	9.5	3.2	10	gen
(ex Skua-92, Union Gem-89)								
Hibiscus Maritime Ltd.								
SEAGULL II (SV)	77	721	1010	60.0	9.5	3.2	10	gen
(ex Osprey-92, Union Saturn-89)								
Cactus Ltd.								
SNIPE (SV)	77	721	1010	60.0	9.6	3.2	10	gen
(ex Petrel-92, Union Pearl-89)								
Daisy Ltd.								
TERN (SV)	77	721	1018	60.0	9.5	3.2	10	gen
(ex Prion-92, Union Jupiter-89)								

SEAGULL II inward bound in the River Ouse *(David H. Smith)*

TANKER TRANSPORT SERVICES N.V., Rotterdam

Funnel: *White with black top and blue-bordered red band bearing a white disk with a blue device consisting of a stylised bird, ship's bow and globe.*
Hull: *Orange with brown boot topping.*
Managers for:

Crestar Shipping B.V.

CRESTAR	81	2538	2550	80.9	13.0	5.2	12	oil/ch/bit
(ex Keke-90, Torasund-85)								

THEODORA TANKERS B.V., Rotterdam

Funnel: *White with two green swallow-tail flags.*
Hull: *Grey with red boot topping.*

STELLA LYRA	89	2874	3480	95.8	14.5	5.7	12	tk
STELLA ORION	73	1600	4382	82.9	13.6	5.8	12	tk
STELLA POLLUX	81	2523	4200	93.1	13.5	5.8	12	tk
STELLA PROCYON	78	2697	4520	83.6	15.6	6.7	12	tk
THEODORA	91	4098	6616	110.6	17.0	7.1	14	tk

The coastal tanker STELLA POLLUX approaches Eastham and the entrance to the Manchester Ship Canal *(Mike Tomlinson)*

TRINITAS SCHEEPVAART KANTOOR B.V., Rotterdam

Managers for:

Majorat ShippingCo. Ltd

TRINKET (Cy)	91	1574	1890	81.2	11.3	3.7	10	gen (72c)
(ex Nessand-94, Hanse Controller-91)								

UNITY CHARTERING (UCS) B.V., Ridderkerk

Managers for:

Rederij Loosman en Ten Rapel

DINA JACOBA	77	987	1473	64.4	10.7	4.0	11	gen
(ex Anne-87, Dependent-82)								

V.L. VAN DER EB, Rotterdam

Funnel: *Yellow with blue band and houseflag which is red with a white diamond and blue letters IS (charter to International Shipbrokers Ltd., London)*
Hull: *Grey with black boot topping.*
Managers for:

B.V. Zeelichter "Thames II"

WILHELMINA V	75	959	1450	65.0	10.8	4.1	11	gen

WILHELMINA-V in the Solent *(Chris Bancroft)*

VAN UDEN MARITIME B.V., Rotterdam

Funnel: *Yellow with blue panel carrying white diamond with letter U in blue.*
Hull: *Red with green boot topping.*
Managers for:

Corpesco Shipping N.V.

BONAIRE (NA)	78	3456	3482	80.2	16.1	6.0	12	gen (165c)

(ex Lekhaven-94, Slotergracht-93, Westafcarrier-79)
Agents only for:

Krony Shipping & Trading Co. S.A.

KLAAS I (Ho)	71	980	1392	70.3	9.9	3.8	10	gen

(ex Willem B-90, Stephan J-80)

ANTHONY VEDER & Co. B.V., Rotterdam

Funnel: *Charterer's makings, commonly Unigas International: blue with red band bearing letter U in blue, with red flame on blue top.*
Hull: *Orange-red.*

CORAL ACROPORA	93	2274	1800	77.4	14.0	5.1	12	lpg
CORAL ACTINIA	93			101.4	14.0	5.1	12	lpg

(recently lengthened)

CORAL ANTILLARUM	82	2400	3030	77.0	14.5	6.2	11	lpg
(ex Black Star-95, Ibiza Star-92, Quevedo-89)								
PRINS ALEXANDER	85	1552	1643	64.3	13.7	4.9	11	lpg
PRINS JOHAN WILLEM FRISO	89	3862	4905	97.4	15.9	6.0	14	lpg
PRINS PHILIPS WILLEM	85	1552	1648	64.2	13.7	4.9	11	lpg
PRINS WILLEM II	85	1552	1648	64.2	13.6	4.9	11	lpg
Newbuildings:								
CORAL OBELIA	96	3835	4258					lpg
CORAL MEANDRA	96	3835	4258					lpg
CORAL MILLEPORA	96	3835	4258					lpg

Also larger ships.

Inward bound on the River Tees, the gas tanker PRINS WILLEM II carries Unigas funnel colours *(Richard Potter)*

VERTOM SCHEEPVAART EN HANDELSMAATSCHAPPIJ B.V., Rotterdam

Funnel: *Blue or green with white disc bearing letter V in green.*
Hull: *Blue or green with red boot topping.*
Managers for:

Rederij G.S. Joustra								
AGNES	78	942	1559	65.8	11.1	4.3	11	gen (54c)
(ex Expansa II-85)								
Twislock Pte. Ltd.								
EXPLORER (Sg)	84	3949	6025	106.6	18.1	6.6	14	gen (419c)
(ex Frisian Explorer-87, Samsun Express-85, Carina Smits-84)								
Lady Anna Shipping Co. Ltd.								
LADY ANNA (Cy)	89	2351	3200	87.0	13.0	5.1	11	gen (153c)
(ex Amrum-95, Port Sado-93)								
Lady Clara Shipping Co. Ltd.								
LADY CLARA (Cy)	90	2351	3200	87.0	13.0	4.5	11	gen (153c)
(ex Baltrum-95, Port Faro-93)								
Lady Greta Shipping Co. Ltd.								
LADY GRETA (Cy)	89	2351	3200	87.0	13.0	5.1	11	gen (153c)
(ex Borkum-95, Port Vouga-93)								

Lady Linda Shipping Co. Ltd.								
LADY LINDA (Cy)	89	2351	3200	87.0	13.0	5.1	11	gen (153c)
(ex Mellum-94, Port Lima-93)								
Lady Lisa Shipping Co. Ltd.								
LADY LISA (Cy)	90	2351	3200	87.0	13.2	5.1	11	gen (153c)
(ex Rottum-94, Port Foz-93)								
Canadel Marine Ltd., Limassol								
LEE FRANCES (Cy)	85	3307	3980	92.2	15.9	6.6	14	gen (217c)
Saltra Chartering & Agencies B.V.								
SALINE	93	1990	3604	89.9	13.0	5.0	11	gen
(ex Salt Trader-93)								
Pearlbay Shipping Co. Ltd.								
SUNSHINE PEARL (Cy)	79	3353	1770	93.8	18.4	3.5	15	ro (52c)
(ex Balder Haren-87, Arena-86, Balder Haren-86, Balder Maracai-81, Balder Haren-79)								
Crodane Pte. Ltd.								
TIGER STREAM (Sg)	85	3949	6025	106.6	18.1	6.6	14	gen (419c)
(ex Faith-96, Frisian Faith-87, Akak Faith-86, Frisian Faith-86, Esa 1-85, Frisian Faith-85, Samsun Faith-85)								

The Cyprus-flag LEE FRANCES passes Hoek van Holland with a cargo of containers *(Roy Fenton)*

VROON B.V., Breskens

Funnel: *White with dark blue or black top with blue letter V breaking three waves, all in light blue. Livestock carriers: white, blue or black top and blue device consisting of letters L and E.*
Hull: *Grey with red boot topping.*
Managers for:

Bonita Marine Management Inc.								
ANGUS EXPRESS (Pl)	67	990	957	80.8	11.0	3.4	14	l/v
(ex Mediterranean Express-87, Amstelstroom-75)								
CHAROLAIS EXPRESS (Pl)	62	3410	1440	108.3	13.2	5.3	16	l/v
(ex Marie Therese le Borgne-72)								
SAHIWAL EXPRESS (Pl)	69	2189	878	76.4	13.4	4.2	14	l/v
(ex El Malek Faisal-81, Cabries-77)								
Maunlad Navigation Inc.								
BRAHMAN EXPRESS (Pl)	66	554	931	80.8	10.9	3.4	12	l/v
(ex Car Express-81, Rijnstroom-76)								

Lawin Maritime Corp.								
BUFFALO EXPRESS (Pl)	83	2374	1600	81.8	14.0	4.1	12	l/v
Motorschip Cold Express B.V.								
COLD EXPRESS (Va)	79	2768	3443	97.0	14.0	5.2	14	ref
Philippine Pacific Ocean Lines Inc.								
GALLOWAY EXPRESS (Pl)	60	3442	3130	119.5	15.0	6.4	15	l/v
(ex European Express-77, Cap Ivi-76, European Express-76, Ladon-74, Cylon-60)								
GUERNSEY EXPRESS (Pl)	67	4255	1358	92.0	17.4	4.7	12	l/v
(ex Viking IV-81)								
Motorschip "Ice Express" B.V.								
ICE EXPRESS (Va)	78	2768	3387	97.0	14.0	5.1	14	ref
Baffin Maritime S.A.								
IRISH PROVIDER (Pa)	64	1466	899	75.8	10.6	3.8	13	l/v
(ex Lidrott-81, Haukeli-79, Tor Flandria-72, Flandria-71)								
Graz Maritime Inc.								
IRISH ROSE (Pa)	65	1466	836	75.7	10.6	3.8	13	l/v
(ex Leo-81, Tor Brabantia-71, Brabantia-71)								
Katingang Shipping Corp.								
NEPTUNIC (Pl)	89	3975	5165	109.9	16.3	7.4	16	ref (82c)
Ideal Maritime Corp.								
NORTHERN EXPRESS (Pl)	86	3978	5175	109.1	16.3	7.3	16	ref (93c)
m.s. Northern Explorer B.V.								
NORTHERN EXPLORER	91	3999	5129	109.9	16.3	7.3	16	ref (20c)
Motorschip Zebu Express B.V.								
ZEBU EXPRESS (Pl)	84	2513	1600	81.7	14.0	4.1	12	l/v

Also larger ships.

WAGENBORG SHIPPING B.V., Delfzijl

Funnel: *Black with two white bands. Managed vessels carry owners' funnel markings but usually fly Wagenborg's houseflag, which is quartered diagonally red and white with a reproduction of the funnel.*
Hull: *Grey with broad orange band with diagonal stripes and WAGENBORG in white and black or red boot topping. Managed ships: Various.*
All the following vessels are managed by Wagenborg for an individual owning company:

AMSTELBORG	78	1575	1500	84.3	10.8	3.8	10	gen (63c)
(ex Rhein-90, Rheintal-88)								

AMSTELBORG at Ipswich: note Wagenborg's flag at the foremast *(Barry Standerline)*

BALTICBORG	91	1999	3015	82.0	12.6	4.9	11	gen (128c)
BOTHNIABORG	91	1999	3005	82.0	12.6	4.9	11	gen (128c)
EEMSBORG	90	1999	3015	82.0	12.6	4.9	12	gen (128c)
FLINTERBORG	90	1999	3015	82.0	12.6	4.9	12	gen (128c)
GAASTBORG	96	2769	4171	90.3	13.7	5.7	14	gen (211c)
GEULBORG	94	2769	4200	90.2	13.7	5.7	12	gen (211c)
GOUWEBORG	94	2769	4200	90.2	13.7	5.7	14	gen (211c)
GRIFTBORG	95	2771	4140	87.7	13.7	5.7	13	gen (211c)
RIJNBORG	91	1999	2952	82.1	12.6	4.9	11	gen (128c)
SCHELDEBORG	91	1999	3030	82.0	12.6	4.9	11	gen (128c)
Managers for:								
B.C. Terstege								
AMBASSADEUR	84	998	1490	78.9	10.1	3.1	10	gen
(ex Michel-92)								
Vuurborg Scheepvaart B.V.								
AMY	89	1999	3037	82.0	12.6	5.0	12	gen (128c)
Coastalaria Shipping Ltd.								
ARIA (Cy)	67	686	1153	58.9	9.0	3.8	9	gen
(ex Stortemelk-92, Dolfijn 1-84, Dolfijn-79, Anda-79)								
Straight Shipping Ltd.								
ATOL (Cy)	80	865	1195	62.8	10.7	4.0	12	gen
(ex Urkerland-94, Autol-90, Punta Motela-87)								
Leibulk B.V.								
AVEBE STAR	78	1094	1180	62.9	11.2	3.6	10	starch
(ex Star-85)								
Hartman Shipping B.V.								
BARENTSZZEE	73	1045	1536	71.2	9.8	3.8	10	gen
(ex Gersom-95, Realta-87)								
Orama Shipping Ltd.								
BISCAY (Cy)	82	1576	2530	81.3	12.1	4.4	11	gen
(ex Eems-94, Concord-94, Concordia-88)								
Rederij motorschip Carolina								
CAROLINA	88	851	1278	64.3	10.6	3.4	10	gen
(ex Feran-91)								
Coastalboscoshipping Ltd.								
CARSCO (Cy)	75	1183	1300	74.9	10.2	3.9	10	gen (18c)
(ex Bosco-95, Lisa B-91, Navigare-85)								
Motorschip Christina C.V.								
CHRISTINA	85	1391	2285	80.0	11.1	4.1	9	gen (86c)
Navyleader Shipping Ltd.								
CITO (Cy)	93	1596	2511	81.7	11.1	4.5	14	gen
Citox Shipping Ltd.								
CITOX (Cy)	75	1183	1652	75.0	10.2	3.9	10	gen
(ex Cito-94, Edina-89)								
Coldstream Merchant N.V.								
COLDSTREAM MERCHANT (NA)	89	4059	6000	99.6	17.0	6.5	12	gen/ta (253c)
(ex Skutskär-93)								
Coldstream Shipper N.V.								
COLDSTREAM SHIPPER (NA)	89	4059	5905	99.5	17.0	6.5	12	gen/ta (253c)
(ex Norrsundet-93)								
Coldstream Trader N.V.								
COLDSTREAM TRADER (NA)	90	4059	5905	99.6	17.0	6.5	12	gen/ta (238c)
(ex Aldabi-93)								
H.J. Smith								
DELFBORG	78	3698	5595	83.1	16.7	7.8	12	gen (181c)
MAASBORG	74	2905	3658	81.8	15.3	6.0	12	gen

Hartmann Scheepvaart								
DEO GRATIAS	74	916	1401	79.9	9.0	3.0	10	gen
(ex Deo Volente-95, Maria-92, Pia-82, Cargo-Liner II-81)								
Hartmann Kustvaart								
DEO VOLENTE	80	1513	1725	82.5	11.4	3.6	10	gen (72c)
(ex Elbstrand-95)								
Rederij M.W. Potkamp & Co. C.V.								
EENDRACHT	80	1009	1596	65.8	11.1	4.3	11	gen
Esmeralda Shipping N.V.								
ESMERALDA (NA)	70	2072	2532	81.8	13.1	6.0	13	gen
(ex Rijnborg-89)								
C.V. Scheepvaartonderneming Flinterdam								
FLINTERDAM	95	3177	4506	99.9	13.6	5.6	13	gen (252c)

FLINTERDAM in the Kiel Canal *(Dominic McCall)*

C.V. Scheepvaartonderneming Flinterdijk								
FLINTERDIJK	78	1863	2955	80.2	12.5	5.5	12	gen
(ex Maritta Johanna-92)								
C.V. Scheepvaartonderneming Flinterland								
FLINTERLAND	94	2818	4216	91.5	13.6	5.8	12	gen (245c)
Flintermar C.V.								
FLINTERMAR	94	2818	4216	91.5	13.6	5.8	12	gen (245c)
c.v. Scheepvaartonderneming Futura								
FUTURA	95	1682	2500	81.7	11.0	4.5	12	gen
Galatea C.V.								
GALATEA	85	998	1490	78.0	10.6	3.2	10	gen
(ex Galaxa-92)								
A.G.M. Peperkamp								
GELRE	92	1595	2168	81.7	11.0	4.1	12	gen
Spiliada Maritime Co. Ltd.								
HYDRA (Cy)	79	1545	1595	85.0	10.0	3.7	10	gen
(ex Waalborg-93, Vera Rambow-89)								
Linde Lloyd B.V.								
IJSSELLAND	85	3757	6446	96.8	15.5	7.1	12	gen (215c)
(ex Lauwersborg-85)								
LENNEBORG	83	3222	5412	82.4	15.8	7.5	11	gen (146c)
LINDEBORG	82	3074	5327	82.4	15.8	7.5	11	gen (146c)

Van Bruggen Scheepvaart								
JAMIE	76	1086	1549	65.0	10.8	4.3	14	gen
(ex-Marico-95, Gina P-93, Kwintebank-92, Els Teekman-84)								
C.V. Scheepvaartonderneming "Jason"								
JASON	93	1681	2503	81.7	11.0	4.5	12	gen
Motorschip Jehan C.V.								
JEHAN	85	1399	2293	79.8	11.1	4.1	9	gen (86c)
Hakvoort Shipping Ltd.								
LEENDERT SR (Cy)	75	849	1196	73.3	9.4	2.9	10	gen
(ex Ambassadeur-92, Marietje-Benita-91, Elisabeth S-83, Willy-79)								
Sandino Shipping N.V.								
LIBELLE (NA)	76	1000	1519	65.8	10.8	4.3	11	gen
(ex Polaris-89, Emmaplein-81)								
B. Kruizinga								
LIFANA	83	928	1452	79.3	10.4	3.2	10	gen
(ex Thalassa-91)								
Scheepvaartbedrijf "Maas" C.V.								
MAAS	83	1316	2175	73.3	11.4	4.2	9	gen
(ex Stern-95, Maas-93)								
He-Wi Shipping Ltd.								
MAASVALLEI (Cy)	84	998	1498	78.9	10.0	3.5	10	gen
H.J. Danser								
MARIETJE ANDREA	92	1599	2250	82.2	11.0	4.1	12	gen
Martini V.o.f.								
MARTINI	74	777	1191	70.1	9.5	2.9	11	gen
(ex Zodiac-92, Wandelaar-91)								
H. Boeree								
MERCATOR	71	1359	2226	76.3	11.9	3.9	13	gen (136c)
(ex Lautonia-89, Heinrich Knuppel-85)								
Lambertus Switijnk								
MICHEL	93	1576	2246	81.7	11.1	4.1	10	gen (128c)
SKYLGE	91	1276	1686	79.1	10.5	3.7	10	gen
(ex Terschelling-91)								
J. Tuininga								
NES	95	1682	2500	81.7	11.1	4.5	12	gen
(ex Boeran-96)								
Rederij m.s. "Nescio"								
NESCIO	93	1596	2190	81.7	11.0	4.5	11	gen (128c)
H.J. Held								
PANDA	89	852	1260	64.2	10.6	3.4	11	gen
(ex Waran-93)								
Scheepvaartbedrijf Gebr. Waker C.V.								
PIONEER	94	2735	4267	89.6	13.2	5.8	11	gen
ZEUS	92	2000	3030	82.0	12.5	4.9	12	gen (128c)
Pionier Shipping C.V.								
PIONIER	85	1486	2261	79.8	11.1	4.1	9	gen (88c)
Rederij m.s. "Reest"								
REEST	87	920	1450	64.2	10.5	3.7	10	gen
(ex Laura II-94)								
H. Koke								
REGINA	86	851	1276	64.3	10.6	3.3	10	gen
(ex Nes-95, Lingedijk-91)								
D. van Eerden								
SAGITTA	88	851	1276	64.3	10.5	3.4	10	gen
(ex Turan-93, Nescio-93)								
Rederij "De Noord"								
SAMBRE	89	1994	3015	83.0	12.6	5.0	11	gen (128c)

G.J.B. Kalkman								
SANDETTIE	89	852	1279	64.3	10.6	3.4	10	gen
(ex Voran-93)								
Aranoyas Shipping N.V.								
SAYONARA (NA)	86	920	1496	64.2	10.6	3.7	10	gen
(ex Laura-93)								
Buter Shipping Co. C.V.								
SCHOKLAND	86	852	1280	64.3	10.6	3.3	9	gen
(ex Buizerd-96)								
Scheepvaartbedrijf "Scorpio" C.V.								
SCORPIO	84	1171	1586	79.1	10.1	3.4	11	gen
(ex Morgenstond-92)								
Gerhard Ahrens K.G. Reederei motorschiff Magula								
SEA MAGULA (Ge)	80	1655	1548	83.0	11.4	3.2	11	gen (72c)
Reederei Frank Dahl m.s. "Merlan"								
SEA MERLAN (At)								
(ex Merlan-78)	78	1495	1550	76.8	11.5	3.4	11	gen (62c)

Although owned in Germany, SEA MERLAN is managed by Wagenborgs at Delfzijl *(David H. Smith)*

Reederei Frank Dahl m.s. "Orade" K.G.								
SEA ORADE (Ge)	90	1354	1699	77.0	11.4	3.2	10	gen (94c)
(ex Orade-91)								
Satineris Shipping Ltd.								
SIRENITAS (Cy)	77	906	1373	63.0	10.0	3.6	10	gen
(ex Sandettie-93, Espero I-89, River Herald-88, Riosal-83)								
Gerrit Tilma								
SIROCCO	88	851	1280	64.3	10.6	3.4	10	gen
(ex Boeran-91)								
C. Vermeulen								
SOLON	92	1595	2500	79.7	11.0	4.2	12	gen
Motorschip Steady B.V.								
STEADY	93	1597	2190	81.7	11.1	4.1	12	gen
Wietze Woudstra								
STERA	77	681	1067	62.2	9.5	3.0	13	gen
(ex Stern-92, Marco-84)								

Scheepsvaartbedrijf Tendo								
TENDO	95	2050	3370	88.0	12.8	4.9	12	gen
C.V. Scheepvaartonderneming Tim								
TIM	79	825	1151	63.6	9.5	3.4	10	gen
(ex St. Michael-94, Jeanette B-90, St. Michael-89)								
V.o.f. Veninga & Zonen								
VEGA	92	1576	2500	81.7	11.1	4.1	12	gen (128c)
Helmuth & Wilfried Rambow								
VERA RAMBOW (Ge)	91	1559	1850	79.7	11.1	3.7	10	gen (80c)
P. van der Pol								
WATERMAN	81	798	1160	70.0	9.5	2.9	11	gen
(ex Marietje Andrea-92, Grote Beer-81)								
Scheepvaartbedrijf Van Berchum V.o.F.								
WILLEM SR.	83	999	1474	79.9	10.1	3.2	11	gen
(ex Steady-92, Yvonne-89)								
C.V. "Zwartemeer" J.& W. van Veen								
ZWARTEMEER	91	1999	3030	81.5	12.4	5.0	10	gen (128c)

Also larger ships.

REDERIJ WANTIJ B.V., Papendrecht

Funnel: *White with mermaid in yellow and blue.*

DONAU	93	2058	3250	81.1	12.4	5.2	11	gen (96c)

WIJNNE & BARENDS B.V., Delfzijl

Funnel: *Black with white band bearing black letters W&B. Some managed vessels carry owners' markings.*
Hull: *Red with black boot topping.*

MARIE CHRISTINE	91	2561	3284	88.0	12.6	5.3	11	gen (129c)
MATHILDE	95	2561	3332	88.0	12.6	5.3	11	gen (129c)
NORA	82	1999	2954	83.3	12.6	5.4	12	gen

NORA picks up her pilot off the Hoek van Holland *(Jan van der Klooster)*

Managers for:								
H. de Weerd								
ANDRIES	72	648	850	70.0	8.2	0.8	10	gen
(ex Thalassa-83, Douwe-S-81, Trinitas-77)								
Rederij Froma								
BETTY	75	1876	3157	89.6	11.8	5.5	12	gen
(ex Inger-93)								
Rederij m.s. Claudia								
CLAUDIA	83	3259	5050	91.9	15.1	6.5	12	gen
Scheepvaartonderneming Constance B.V.								
CONSTANCE	85	3259	5105	91.9	15.1	6.5	12	gen
G. & P. Schalk								
DEPENDENT	82	1241	1791	65.5	11.4	4.6	11	gen
Rederij Sanitas								
ERNA	79	1599	2822	75.1	12.6	5.4	12	gen
(ex Menna-95)								
C.V. Flardinga Scheepvaartbedrijf Doorduin B.V.								
FLARDINGA	91	1282	1750	75.0	10.8	3.7	10	gen
C.V. m.s. "Helene"								
HELENE	91	2561	3284	88.0	12.6	5.3	12	gen (129c)
Scheepvaartbedrijf J.R. Klein & Co. C.V.								
HENDERIKA KLEIN	83	1865	3077	81.6	14.0	5.4	12	gen
Rederij Isabel B.V.								
ISABEL	72	1472	2159	71.3	11.6	5.0	12	gen
C.V. m.s. Kirsten								
KIRSTEN	95	2561	3290	88.0	12.6	5.3	12	gen (129c)
(ex Aros News-96, Kirsten-95)								
Rederij Linda Marijke								
LINDA MARIJKE	92	1359	1850	75.3	10.8	4.0	10	gen
C.V. m.s. "Magda"								
MAGDA	93	2561	3284	88.0	12.6	5.3	12	gen (129c)
Rederij Margriet								
MARGRIET	77	1114	1570	65.8	10.7	4.3	12	gen
(ex Alecto-89, Marjan-86)								
Rederij C.T. Drent & Zn.								
MORGENSTOND I	93	2650	4250	89.0	13.2	5.7	12	gen (190c)
MORGENSTOND II	90	1999	2950	82.0	12.5	4.9	12	gen (96c)
C.V. m.s. "Olga"								
OLGA	94	2561	3290	88.0	12.6	5.3	12	gen (129c)
Shipping Co. Intertrade S.A.								
SIAN (Ho)	75	1547	2723	73.4	11.8	5.5	14	gen
(ex Irina-91, Coenraad Kuhlman-86)								
Rederij motorschip Symphony								
SYMPHONY (At)	71	1395	2226	76.2	12.1	5.1	13	gen (104c)
(ex Capella-95, Condor-94, Markus-90)								
G. Switijnk								
THALASSA	92	1786	2200	87.0	12.4	3.7	11	gen
Rederij Verena								
VERENA	77	1599	2677	75.1	12.6	5.4	12	gen
(ex Mathilde-94)								

J.L.A.M. WILDENBEEST, Enschede

GEMINUS	71	402	564	55.0	7.2	2.2	10	gen
(ex Zuiderzee-85, Bornrif-83, Veritas-79, Dolfijn-75)								

BELGIAN SHORT SEA SHIPS

AHLERS SHIPPING N.V., Antwerp

Funnel: *Blue with broad white band bearing blue triangular device.*
Hull: *Blue.*
Managers for:

Belgian Shipping N.V.

BRABO (Cy)	84	2859	3600	89.9	15.9	6.6	13	gen (245c)
(ex Norasia Adria-89, Brabo-87)								

Also larger ships.

BIBUTANK S.A., Kalmthout

INNO	73	1607	3150	110.0	11.4	2.8	12	tk
(ex Inez-89)								

B.V.B.A. D.N.R. TANKVAART & AGENTUREN, Ossendrecht

Funnel: *Black with houseflag which is white and red with blue letters D.N.R.; or red with white band and black letters D.N.R.*
Hull: *Various*
Managers for:

Balata Ltd.

GOLDCREST (IOM)	65	575	762	62.3	9.5	3.1	10	ch tk
(ex Silverkestrel-94, Goldcrest-92, Carrick Kestrel-87, Silverkestrel-75)								
SANDLARK (IOM)	66	688	952	57.0	9.8	4.4	11	ch tk
(ex Silverlark-94, Sandlark-92, Ice Lark-87, Finnlark-76)								

Associated company:

B.V.B.A. De Neef Rederij, Antwerp

POLO (At)	73	1648	2350	110.0	11.4	3.4	12	tk
(ex Charlotte F-91, Ricy VI-85)								

The Belgian managed chemical tanker SANDLARK approaches Burton Stather on the River Trent *(David H. Smith)*

DELTA SHIPPING & TRADING B.V.B.A., Antwerp

Funnel: *Blue with white panel containing orange vertical arrow above blue triangle flanked by blue letters D S.*
Hull: *Blue with red boot topping.*
Agents only for:

Golden Deal Shipping Ltd.								
AROSSEL (Cy)	72	2454	2443	88.5	13.9	5.3	14	gen (128c)
(ex Rossella-91, Rodano-85, Progress Link-82, Scol Progress-81)								
Managers for:								
Skipper Marine Ltd.								
BIANCO DANIELSEN (Cy)	76	1836	2350	79.5	13.1	4.8	12	gen
(ex Pia Danielsen-88)								
MARCO DANIELSEN (Cy)	77	1895	2595	79.8	13.1	5.2	12	gen
(ex Ilha do Porto Santo-94, Winni Helleskov-81)								
PIA DANIELSEN (Cy)	78	1906	2535	80.0	13.1	5.2	13	gen
(ex Atlantic Rainbow-93, Anny-88, Anny Danielson-85)								
Golden Venture Shipping Ltd.								
HERMAN BODEWES (Cy)	78	2697	3636	82.2	15.1	6.1	13	gen (60c)
(ex Herman Danielsen-86, Herman Bodewes-84)								
Inter-Laxo Compania Naviera S.A.								
IRENE VI (Pa)	63	497	1152	67.0	10.5	3.8	11	gen
(ex Ursula L-83, Lieselotte H-78, Luhe-74, Visurgis-70)								
Waterdrive Marine Ltd.								
ULLA (Cy)	81	2704	2700	77.3	15.0	5.9	13	gen
(ex Magnus S-92, Ringvoll-88, Namdalingen-84)								

EURO MARINE SERVICES B.V.B.A., Essen

Funnel: *Various.*
Hull: *Various, often blue.*
Agents for:

Urkerland Shipping Ltd.								
BATAVIER (Cy)	86	1543	2200	87.6	11.1	3.9	10	gen (98c)
(ex Venus-94, Bromley Sapphire-92, Union Sapphire-90)								
Newtrend Shipping Ltd.								
BREEZAND (Cy)	83	1560	2401	87.8	11.1	4.1	9	gen (98c)
Pelops Shipping Ltd.								
CAST BASS (Cy)								
(ex Mirka, Nevskiy 37-92)	92	2882	2250	108.9	15.5	2.8	11	gen (200c)
Marine Delta Shipping Ltd.								
CAST SALMON (Cy)								
(ex Nevskiy 36-92)	92	2691	2845	108.9	15.5	3.2	11	gen (200c)
Juno Shipping Ltd.								
EMS-LINER (Cy)	77	1300	1450	84.8	9.5	3.3	11	gen
Seaconqueror Shipping Ltd.								
LLANO (Cy)	81	1010	1448	70.0	11.3	3.4	11	gen
(ex Union Pluto-94)								
Mediamond Navigation Ltd.								
MANJA (Cy)	74	935	1476	80.0	9.0	3.2	9	gen
(ex Uelzen-82, Cargo-Liner IV-81)								
V.o.f. Paphos								
PAPHOS (Du)	92	1999	3186	81.2	12.4	5.2	11	gen (125c)
Sealotus Shipping Ltd.								
WATERWAY (Cy)	81	1010	1448	69.9	11.3	3.4	11	gen
(ex Union Venus-94)								

The Russian built CAST SALMON leaves the Nieuwe Waterweg on one of her regular voyages from Dordrecht to Zeebrugge with containers
(Jan van der Klooster)

Registered in Cyprus but Belgian managed, EMS-LINER passes Reedness outward bound from Goole *(David H. Smith)*

FOKA-GAS N.V., Kapellen

Funnel: *Black with white letters FK. Santona Shipping: blue with orange letter S.*
Hull: *Black or orange.*

FOKA-GAS 1	67	1216	1152	88.1	11.5	2.5	9	lpg
(ex Ricy Gas-87, Sarriette-77, Romarin-76)								
Managers for:								
Santona Shipping Co. Ltd.								
QUEENY MARGRETH (Cy)	74	1202	950	67.9	11.2	4.5	12	lpg
(ex Shogen Maru-91)								

KILGAS PIONEER arrives in the Bristol Channel to load liquified gas for Drogheda *(Bernard McCall)*

SEATRADE SHIPPING N.V., Essen
SEATREND SHIPPING N.V., Essen

Funnel: *Various.*
Hull: *Various.*

Managers for:								
Oceanlook Shipping Ltd.								
ELKA THERESA (Cy)	90	1468	2440	79.7	10.9	4.2	11	tk
Seaglobal Shipping Ltd.								
KILGAS PIONEER (Cy)	92	1173	1400	76.1	11.4	3.3	10	lpg
(ex Annagas-92)								

Also larger ships.

REDERIJ GUY SOMERS, Antwerp

Funnel: *Yellow with black top and houseflag as band. White with red edges top and bottom with green diamond bearing a white letter S.*

Managers for:

Gemma Shipping Ltd.								
MANUELLA (Cy)	88	1853	3200	109.7	11.4	3.9	11	ch tk
Marinetrend Shipping Ltd.								
SAM (Cy)								
(ex Inez III-95)	85	2054	3720	110.0	11.4	4.0	11	ch tk

STEVEDORING & TRANSPORT N.V., Antwerp

Funnel: *Black with white letter V inside white O (Phs. Van Ommeren).*
Hull: *Blue above black with red boot topping.*

ANTWERPEN								
(ex Lutjenbuttel-79)	79	1064	2249	107.8	9.5	3.4	10	ch tk
GENT								
(ex Jan-87)	79	1471	2340	85.9	11.3	4.0	10	ch tk

BIANCO DANIELSEN hurries up the Bristol Channel on her way to Sharpness. *(Bernard McCall)*

ABBREVIATIONS FOR VESSEL TYPES

aux	auxiliary motor ship
cc (c)	container carrier (container capacity in Twenty foot Equivalent Units)
cem	bulk cement carrier
ch tk	chemical tanker
gen	general cargo
gen (c)	general cargo (container capacity in Twenty foot Equivalent Units)
gen/ta	general cargo with tank capacity
hl	heavy lift ship
lpg	liquified gas tanker
l/v	livestock carrier
oil/ch/bit	oil/chemical/bitumen tanker
ref	refrigerated cargo
ro	cargo ro-ro
ro/gen	ro-ro/general cargo
starch	starch powder tanker
tk	tanker
u	capacity in 12-metre trailer units
veh	vehicle carrier

FLAG ABBREVIATIONS

At	Antigua and Barbuda
Be	Belgium (in Dutch section)
Bs	Bahamas
Bz	Belize
Cy	Cyprus
Du	Netherlands (in Belgian section)
Ge	Germany
Ho	Honduras
IOM	Isle of Man
Le	Lebanon
Ma	Malta
NA	Netherlands Antilles
Pa	Panama
Pl	Philippines
Sg	Singapore
SV	St. Vincent and Grenadines
Va	Vanuatu

Where no flag is shown the vessel is Dutch in the Dutch section and Belgian in the Belgian section.

SHIP NAME INDEX